Think it out

1

Problem-solving exercises in maths

Nigel Peace

EDWARD ARNOLD

© Nigel Peace 1986

First published in Great Britain 1986 by
Edward Arnold (Publishers) Ltd, 41 Bedford Square, London WC1B 3DQ

Edward Arnold (Australia) Pty Ltd, 80 Waverley Road, Caulfield East,
Victoria 3145, Australia

Reprinted 1987

British Cataloguing in Publication Data
Peace, Nigel
    Think it out.
    Bk. 1
    1. Mathematics–Examinations, questions, etc.
    I. Title
    510'.76   QA43

ISBN 0–7131–8407–8

Text set in 10/12 IBM Press Roman
by TecSet Ltd, Sutton, Surrey
Printed by The Bath Press, Avon
Bound by W H Ware & Sons Ltd, Clevedon, Avon

Contents

Acknowledgements

My grateful thanks to Jeffrey for his advice and encouragement, and to all my pupils who not only made this book possible but had to test out all the exercises too!

The cartoons and illustrations have been done by Sam Jacob, Alex Garland, Peter Cackett and Owen Newman, all pupils of University College School, London, and by the author.

1 X-number puzzle 1

Copy the grid carefully.

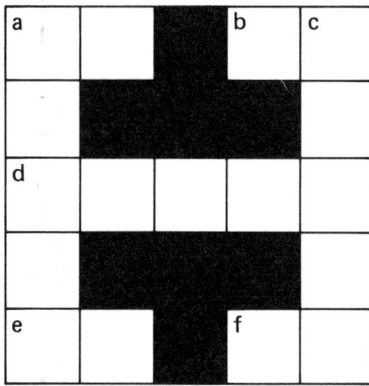

Clues across
(a) 6 + 13 + 9 − 18 − 7 + 9
(b) treble 17
(d) five sixes minus a countdown!
(e) **b** across minus **a** across
(f) one fifth of £2 in pence

Clues down
(a) ten thousand, one hundred and three
(c) 500 × 21

Can you find five different digits (0–9) to fit the grid opposite so that the total of the four outside numbers equals the middle number?

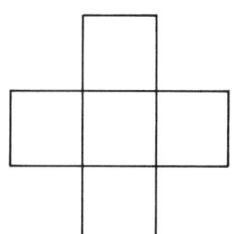

Now can you arrange five different digits so that the total vertically is the same as the horizontal total?

Make your own puzzle
Copy the grid at the top again and put a number in each square. Now make up clues for each line and each column... Set the puzzle for your neighbour and see if you can finish theirs before they finish yours.

2 Amanda's wardrobe

★

Amanda's father said he would make her a wardrobe if she could design one that cost less than £100 – the old meanie! Her basic design is shown here – there are also two doors of equal size, but there is *no* back. The wood costs 75p per square foot.

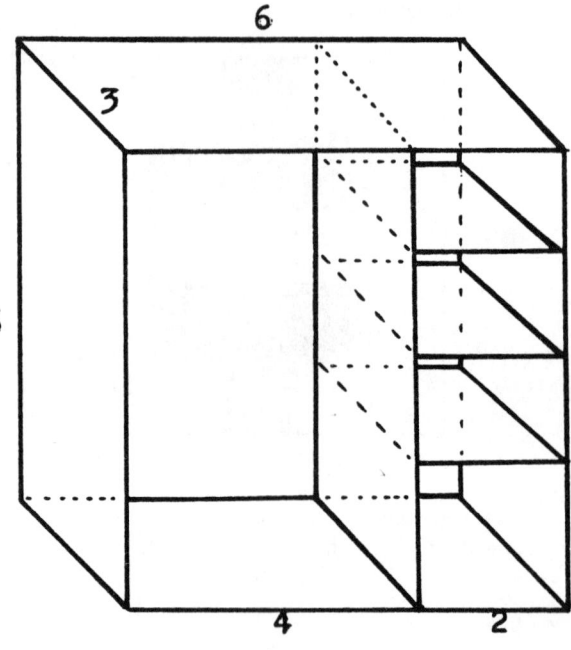

All the lengths are in feet. Copy and complete the list below:

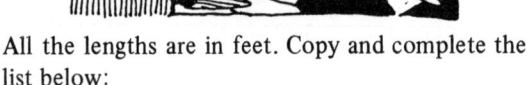

Item	Size	Area	Cost
Side	6 x 3	18 sq. ft	18 x 75p = £13.50
Side	6 x 3		
Top			
Base			
Door			
Door			
Partition			
Shelf		6 sq. ft	
Shelf			
Shelf			
Screws and hinges			£ 2.00
		Total cost	£

You should find that the wardrobe was too expensive! By how much?

Then Amanda had a good idea. She changed the design so that the partition was only 1 foot deep. (It was still 6 feet high, and the shelves were still 2 feet wide.)

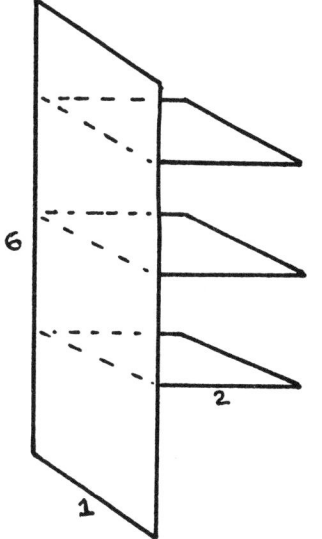

Make a new list and work out the cost of the new design.

Her father was so pleased that he made her a matching bedside table too (shown on the right). Make a list like the one before and find the cost of the table. (Allow £1 for screws.) There are no doors, but there is a back.

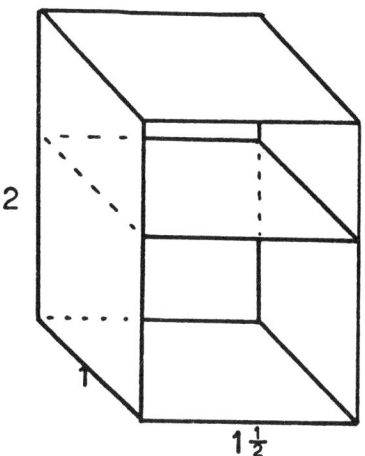

3 The sponsored walk

★

Fifty children went on a sponsored walk to raise money for guide dogs. Twenty of them took their own dogs along too! They walked round a lake; one lap round was 3 miles.

	Children	Dogs	Laps done	Miles done	Children × Miles
	10	6	1	3	10 × 3 = 30
	22	12	2	6	
	8	1	3		
		0			72
	4		5		
Total:	50	20	—	—	

The table shows how they got on. Copy it and fill it in.

(a) How many legs were walking altogether?
(b) How many dogs walked 15 miles?
(c) How many children walked at least 6 miles?
(d) How many children walked at least 10 miles?
(e) Divide the number of legs by the number of tails!
(f) How many dogs walked at least this number of miles?
(g) Wasn't that a silly question?
(h) How much money was raised if, on average, every child was sponsored for 20p a mile? (Hint: children × miles × amount per mile)

(i) If it costs £25 to buy a guide dog puppy, how much more money is needed so that they can buy three dogs? (The cost of training the puppies is far more than this.)
(j) Write down five interesting ideas for sponsored events that your school might hold. What would you like to raise money for?

4 Logic puzzles

Draw 9 crosses in a pattern as shown. Now, can you draw 4 straight lines to pass through all 9 crosses without lifting your pencil from the paper?

x   x   x

x   x   x

x   x   x

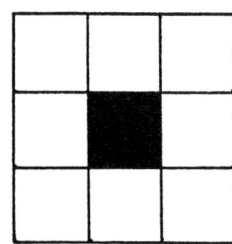

Can you write eight different numbers between 0 and 10 in the squares so that every line adds up to fifteen?

Find the missing numbers here:

$$1, 3, 7, -, 31, 63, -, 255$$

Now find the missing letters:

$$A, C, B, D, -, -, -, F$$

If 4 labourers can dig a hole in half an hour while their 2 mates lean on a wall, how long would it take 4 labourers to lean on the wall while the other 2 dig a hole?

In the game of dominoes, each domino has 2 numbers on it between 0 and 6. For example: 0-5, 3-4, 2-6, 1-1, etc. Each pair of numbers occurs only once in a set. How many dominoes must there be in a set? How many dots are there altogether in a set (the total of all the numbers)?! (Hint: what is the *average* total for each domino?)

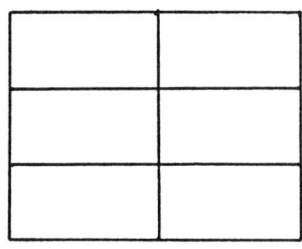

How many rectangles can you find in the diagram on the left? (Is a square a rectangle?)

5 Pop charts

How many of the artists in the Top Twenty Chart to the right have you heard of?

The numbers in brackets show how many places they've moved since the previous week. For example, Paul Young has gone up 6 (+6), so he was at number 13 the previous week.

1	Madness	(+1)
2	Culture Club	(−1)
3	Kajagoogoo	(+2)
4	Duran Duran	(−)
5=	UB 40	(−)
5=	Phil Collins	(−2)
7	Paul Young	(+6)
8	Tracey Ullman	(−)
9	Genesis	(+10)
10	Rod Stewart	(−3)
11	The Beatles	(−2)
12	Simple Minds	(+8)
13	The Cure	(−3)
14	KC And The Sunshire Band	(−3)
15	Bonnie Tyler	(−)
16	Mike Oldfield	(−)
17=	Tina Turner	(−)
17=	Human League	(+1)
19	Roberta Flack	(−5)
20	Howard Jones	(−8)

Write the previous week's chart in full.

The week after Madness were at Number One, the Top Ten was:

1	Duran Duran	(+3)
2	Paul Young	( )
3	Madness	( )
4	Simple Minds	( )
5	Genesis	( )
6	Culture Club	( )
7	Human League	( )
8	UB 40	( )
9	Kajagoogoo	( )
10	Rod Stewart	( )

Copy this and put the correct numbers in the brackets.

When you play a 'single' it revolves 45 times a minute. A record usually revolves for about 15 seconds before the music starts (called 'lead in') and 15 seconds after it ends (the 'run out'). How many times will **(a)** a 3-minute song, and **(b)** a 4-minute song, revolve when you play them?

On the diagram below, the needle starts at A and lifts off at D. The music starts at B and stops at C. From A to B and from C to D there are 10 grooves per centimetre, and from B to C there are 20 grooves per centimetre. A–B and C–D are both 0.5 cm, and B–C is 7.0 cm.

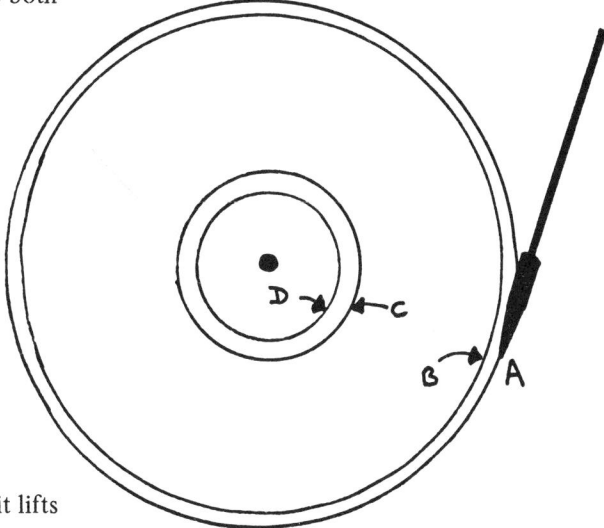

How far does the needle travel from A until it lifts off?

6 Logic 1

★

Put these operations in a sensible order. There may
well be more than one way of doing it – can you
find any variations?

Buying a record

(a) Check you have enough money.

(b) Leave the shop.

(c) Select a record.

(d) Look at the records in the shop.

(e) Go into the shop.

(f) Check the price of the record you want.

(g) Pay for the record.

Getting off to school

(a) Catch the bus.

(b) Put on your coat.

(c) Have breakfast.

(d) Arrive at school.

(e) Wake up.

(f) Get dressed.

(g) Check your school books.

(h) Clean your teeth.

(i) Leave home.

(j) Have a wash.

Making a cup of tea

(a) Pour the tea into the cup.

(b) Switch off the kettle.

(c) Get a cup and saucer out.

(d) Pour boiling water into the teapot.

(e) Make sure there's enough water in the kettle.

(f) Put the teacosy on the teapot.

(g) Switch on the kettle.

(h) Get the teapot out.

(i) Put milk and sugar in the cup.

(j) Put tea in the teapot.

(k) Allow the tea to brew.

7 Money trouble

★

Add up the shopping bills below. How much change will there be from £10 for each one?

Record	£1.50
Book	£2
Pens	50p
Magazine	£1
Photo Album	£3.50
TOTAL	

GROCERY	£3.75
WINE	£2.75
COLA	75p
CRISPS	50p
FAGS	£1.25

THANK YOU!
PLEASE CALL AGAIN.

D.I.Y. STORES LTD

PLYWOOD	£4.60	VARNISH	78p
SCREWS	37p	HANDLE	9p
HINGES	29p	V.A.T.	
GLUE	£1.03	@ 15%	£1.08

I spilled coffee on the two bills shown here! Can you work out how much the missing amounts were?

Coffee	£1.10
Eggs	
Fish	78p
Total	£4.50

Pencil	9p
Compass	49p
Pen set	£3.02
Rubber	8p
Magazines	£1.12
Books	£2.19
	£7.60

At the start of the month I had £10.50 in the bank. During the month I paid £85 rent, £16.50 rates, £22.25 for electricity, £13.40 for telephone calls, £68.32 for food, £8 for fares and £21.25 for other things. At the end of the month my salary of £250.77 was paid into the bank, but the bank charged me £5.75 for services (being overdrawn!). How much have I got in the bank at the start of the next month?

How much better off am I than last mon

8 Amazing!

Here is how to invent a maze. Draw a large grid (say, 10 × 10 squares) and mark a Start and Finish.

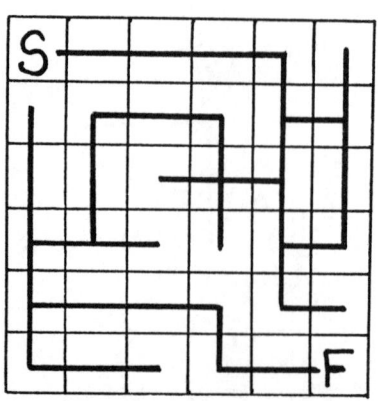

Now, starting at S draw a network of paths going through every square. Try to make the route from S to F difficult, and put in some dead ends!

All the lines that do *not* have a path crossing them now are the *walls* of the maze.

Draw these in clearly, in a different colour if you have one.

Finally, using a new piece of paper copy or trace your maze but draw *only* the walls, and the Start and Finish.

Swop it with your neighbour and see who can find the path through the other's maze first.

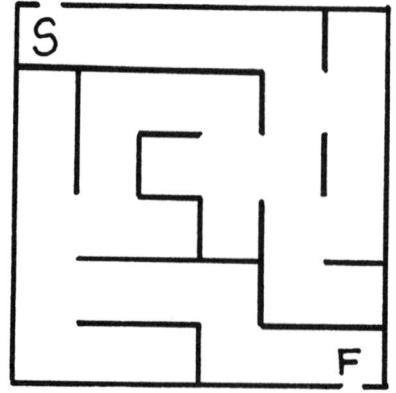

If you feel really clever, now draw some concentric and equally spaced *circles* with straight lines radiating out from the centre, and make a circular maze!

Can you invent some other interesting shapes for mazes?

9 Logic puzzles 2

★

Put the numbers 1, 2 and 3 at the vertices of a triangle. Now try to put 4, 5, 6, 7, 8 and 9 along the sides so that each side adds up to 17.

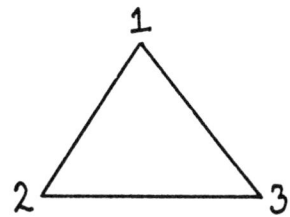

Now arrange the numbers 1–9 round another triangle so that each side equals 20.

How many triangles are there in this design?

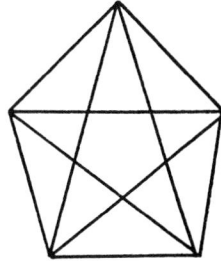

As I was going to St Ives
I met a man with 7 wives.
each wife had 7 cats,
each cat had 7 kittens,
each kitten had 7 fleas,
each flea had 7 microbes.
How many were going to St Ives?!

There are 5 chocolates in a box. How can you share them between 5 children so that each child gets a chocolate but one stays in the box?!

The grid shows a garden divided into plots. A lazy gardener wants to start at X and go through each plot *once* only, returning to X. (The black squares are walls.)

Can you find his route? (Copy the diagram, so that you don't have to write in this book.)

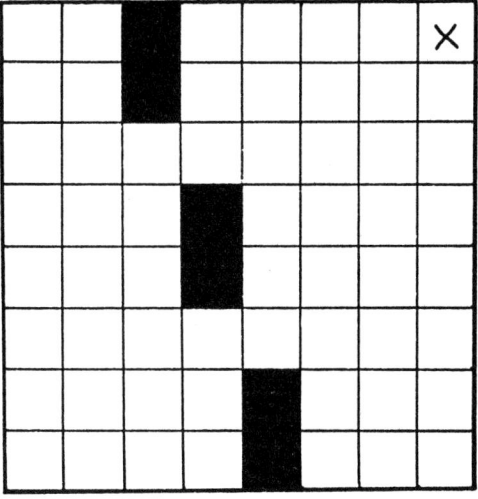

10 Logic puzzles 3
★

If a clock takes 30 seconds to strike 6 o'clock, how long will it take to strike midnight? (The answer is *not* 60 seconds!)

How many years are there between the last year that was the same upside down and the next year that will be the same upside down?!

When I was 6 my father was 30, but he's twice as old as me now. How old am I?

I wanted to make a Christmas decoration in a mathematical pattern, but I only had ten lights. After thinking for a while I made a pattern which had five rows, and each row contained four lights! How did I do it?

Can you find *three* numbers that make the same total when they're added as when they're multiplied together?

Example: $2 + 2 = 2 \times 2$

Can you do the same for four numbers, and for five numbers?

Surely this problem is impossible! Can you find two whole numbers that make 37 when multiplied together??

11 Crowd trouble!

Below are the results in the Football League 1st Division one Saturday a few years ago – times have changed. The attendance at each match is in brackets.

Team	Score	Team	Attendance
Arsenal	3 – 1	Manchester Utd	(40 739)
Aston Villa	0 – 3	Liverpool	(40 190)
Bristol City	3 – 1	Newcastle	(17 344)
Everton	2 – 1	Derby	(38 213)
Leicester	0 – 1	West Brom	(14 637)
Man. City	2 – 1	Ipswich	(34 975)
Norwich	3 – 0	Leeds	(19 615)
Notts Forest	3 – 1	Chelsea	(31 262)
QPR	1 – 0	Middlesborough	(12 925)
West Ham	2 – 1	Coventry	(19 260)
Wolves	0 – 1	Birmingham	(19 900)

(a) Round off the attendances to the *nearest hundred*.

(b) Find the total of these rounded off numbers.

(c) Find the total of the *actual* attendances, and compare this with your last answer – is it surprising?

(d) Now round off the attendances to the nearest *thousand*.

(e) Find the total of these and compare it with the actual total attendance. How close are you this time?

Rounding off numbers often gives a good estimate for a sum and saves a lot of trouble and time. But we have to be careful not to round off too much! Make up 5 sums like the example below, and find their real answers. Then do them against by rounding off to the nearest 10p, then the nearest 50p.

Example:
£2.40 + £1.18 + £3.86 + 96p + £1.08 = £9.48
Nearest 10p:
£2.40 + £1.20 + £3.90 + £1.00 + £1.00 = £9.60
Nearest 50p:
£2.50 + +1.00 + £4.00 + £1.00 + £1.00 = £9.50!

12 Logic puzzles 4

Can you rearrange these anagrams to make mathematical words?

RAQUES

GITHEH

ENCO

LICCER

GLANE

SHEPER

It is midnight. Raining again! Will it be sunny in 72 hours?

One evening all the lights failed as I was getting ready to go out. I found my three pairs of shoes in a pile, and my 12 socks (3 black pairs, 3 blue pairs) in another pile. But I couldn't see any of them properly! How many shoes and socks did I have to take to be sure of having a pair of matching shoes and a pair of matching socks?

The diagram shows 4 dominoes put together to make a sum:

$551 \times 4 = 2204$

Try to arrange other dominoes in sets of four to make other multiplication sums.

If you're *really* clever, make 7 sums like this and use up all 28 dominoes in a set!

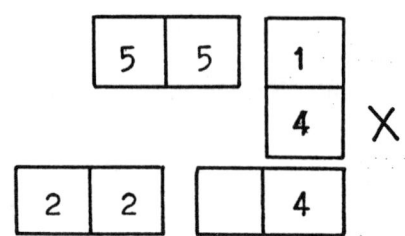

A	B	C	D	E
F	G	H	I	J
K	L	M	N	O
P	Q	R	S	T
U	V	W	X	Y
Z	1	2	3	4
5	6	7	8	9

The grid on the left is the key to a secret code. For example: A is coded 17 (1st column from left, 7th row up); B is coded 27; T is coded 54; 5 is 11, etc.

Decode this:
5355131734573446263654 !
Now code a message of your own and try it on your neighbour.

13 Logic puzzles 5

★

At what time do the hands of a clock divide the face into two parts so that the numbers in each part have the same total? (Two possible answers.)

A boy has the same number of brothers as sisters. But each sister only has half as many sisters as brothers. How many children in the family?

There are twelve trees along my street, spaced out equally with one at each end. If I run from one end of the street I can reach the eighth tree outside my house in just eight seconds. How long would it take me to run down the whole street?

Can you put some + signs between the numbers of the left of the equation to make it correct?
Now try to do the same for these:

| 1 | 2 | 3 | 4 | 5 | 6 | 7 | 8 | 9 | = | 99 |

| 1 | 2 | 3 | 4 | 5 | 6 | 7 | = | 100 |

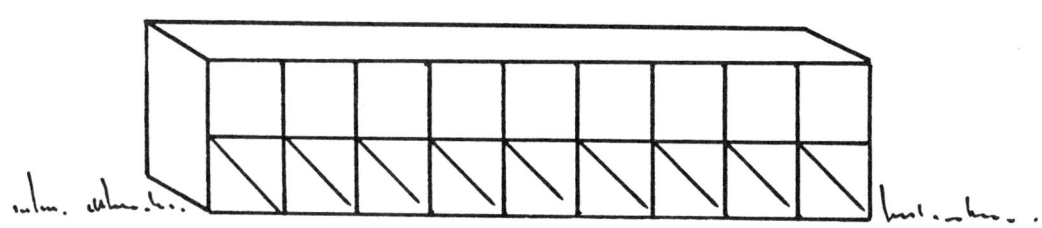

Put ten horses in the nine stables above. Each stable is only big enough for one, and you can't cut the horses up into bits!

14 Design by numbers

★

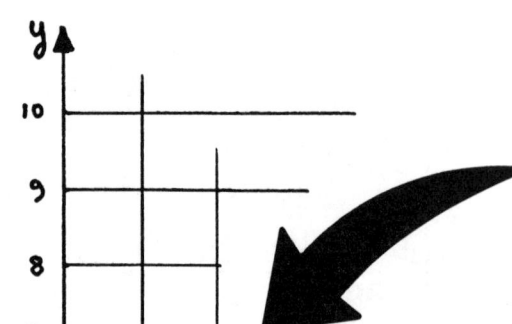

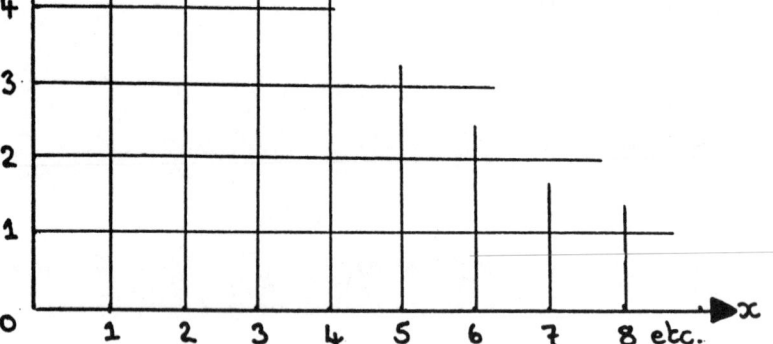

Draw a grid of 25 squares across (*x*) and 10 squares up (*y*). The co-ordinates of a point are the number of squares across from 0, followed by the number of squares up from 0. For example:
This point is (2, 7)

Mark each of the following points and join them up in order: (9, 10) (10, 8) (12, 8) (13, 10) (14, 6) (14, 4) (12, 1) (10, 1) (8, 4) (8, 6).

Now join up (9, 5) (10, 5) (10, 6) (9, 5)
and (10, 2) (12, 2) (22, 3) (10, 2)
and (12, 5) (13, 5) (12, 6) (12, 5).

On another grid of the same size, mark these and join them up: (5, 5) (5, 6) (6, 6) (8, 8) (14, 8) (15, 7) (20, 6) (20, 5) (5, 5). Add windows – and wheels!

Now draw a grid of 25 squares in *both* directions, and join these up: (5, 5) (5, 15), (8. 25) (11, 15) (11, 11) (17, 11) (20, 9) (20, 5) (9, 5) (9, 8) (7, 8) (7, 5) (5, 5).

Draw a figure of your own design on a similar grid. Use straight lines only! Make a list of the co-ordinates for your neighbour and see if he or she can draw your design correctly.

15 Searching!

You are an ant standing at corner A of a cube and your dinner is at corner H. However:

1. you can only walk along the edges of the cube;
2. you cannot move upwards;
3. you don't want to walk along any edge more than once.

How many different ways could you get from A to H? List them like this: (1) A–B–C–H etc.

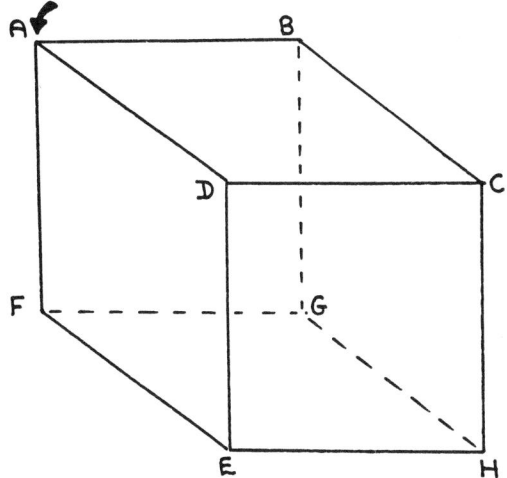

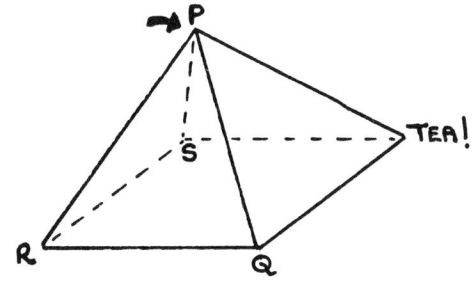

Try the same thing for a square pyramid, a tetrahedron and a triangular prism. Always start at a top corner and finish at a bottom corner.

Copy the word-search and see how many words to do with maths you can find. LINE and ODD have been marked already to get you started.

P	L	A	N	E	R
R	I	A	R	E	A
E	N	S	B	S	G
V	E	M	I	R	P
E	U	D	A	X	D
N	E	P	D	P	A
E	H	E	D	O	C

16 Equation search 1

This puzzle is like a wordsearch, except that you must look for numbers that go together to make equations. Here is an example:

4	5	2	7	3
2	12	9	0	2
11	8	3	29	1

The numbers ringed form equations because

$$5 + 2 = 7$$
$$3 - 2 = 1$$

(There are four more equations like these in the grid on the left!)

Equations may be horizontal, vertical or diagonal.

Copy the grid below and then try to find *thirty* equations in it by using only the signs +, − and =.

5	3	7	10	8	2
7	1	0	20	3	9
6	4	1	15	1	3
6	6	2	5	8	14
5	2	4	10	4	7

17 X-number puzzle 2

Copy the grid carefully first.

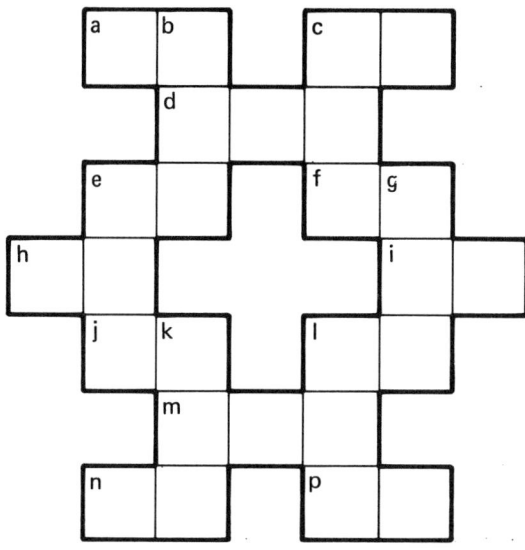

Clues across
(a) $1 \times 2 \times 3 \times 4$
(c) $5 \times 9 - 3$
(d) a gross number!
(e) $5 \times (9 - 3)$
(f) $2 \times 2 \times 3 \times 2 \times 2$
(h) days in June
(i) 20s in a thousand
(j) days in February 1990
(l) days in February 1988
(m) $5 + (5 \times 50) + 6$
(n) you'll be lucky!
(p) the first prime after 47

Clues down
(b) 14 backwards!
(c) three fours
(e) $4 \times 104 - 70 - 44$
(g) $215 \times 4 - 1$
(k) $534 + 289$
(l) **m** across $- 46$

Try to draw a grid of your own in an interesting shape. Put a number in each square, and make up clues.

18 Letumin for the cup!

★★

The Letumin School U13 team had a poor season and, with three games to go, they looked certs for one of the two relegation places. The bottom of the table looked like this:

Team	Played	Won	Drew	Lost	Goals For	Goals Against	Points
Beatumup School	15	6	2	7	14	19	20
Kickem U13	15	6	2	7	14	23	20
Bash Street B	15	5	4	6	16	27	19
Letumin U13	15	3	4	8	9	31	13

Note: **(a)** there are 3 points for a win, 1 point for a draw;

(b) Kickem are below Beatumup because the difference between their Goals For and Goals Against is worse.

But in the last games of the season, Letumin had a sudden change of fortune (new goalie!). These were the results:

Beatumup	0 - 0	Kickem
Bash Street	1 - 1	Beatumup
Bash Street	1 - 1	Kickem
Bash Street	0 - 3	Letumin
Letumin	6 - 0	Kickem
Letumin	2 - 1	Beatumup

Write out the final table after these results.

Example: Beatumup School have played 3 more, won 0, drawn 2, and lost 1. They have scored 2 more goals with 3 more scored against them, and have 2 more points:

Beatumup School 18 6 4 8 16 22 22

Who was relegated after all? Why did this depend on the very last game? What you have happened if this result had been

Letumin   1- 1   Beatumup   ?

In the knock-out cup Letumin did much better.
Copy the chart below and fill in the missing names.
Who won the Cup? Who were runners-up?

Letumin 2	Letumin 7	Letumin 1			
Bash St 0	Toffeenobs 1		 5		
	Beatumup 2	 0			
	Bustagut 3				
	Kickem 1	 4			
	Upwithus 0		 4		
	Cissy Rd 24	 3			
	Notachans 1				

Note that there were 9 teams in the Cup, and 7
had 'byes' in the first round. This was so there
would be a convenient number of 8 for the second
round.

If there had been 12 teams in the Cup, how many
would have to have byes in the first round? How
many if there were 16 taking part?

YOU'LL HAVE TO DO BETTER THAN THAT NEXT WEEK — WE'RE PLAYING IN THE CUP!

19 Everyone in place

The new Maths teacher had a terrible memory for names, so he gave every pupil co-ordinates to indicate where they sat in class. He could remember numbers at least!

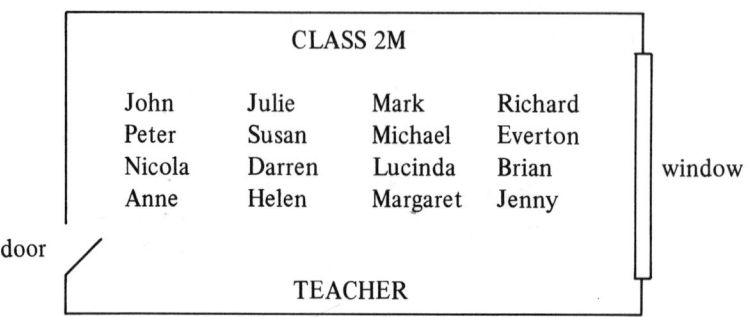

CLASS 2M

John	Julie	Mark	Richard
Peter	Susan	Michael	Everton
Nicola	Darren	Lucinda	Brian
Anne	Helen	Margaret	Jenny

door

window

TEACHER

Example: Helen is (2, 1) because she's 2 seats from the left and 1 from the front. Mark is (3, 4).

Write the co-ordinates for
(a) John
(b) Lucinda
(c) Jenny

Who has these co-ordinates?
(d) (1, 1)
(e) (4, 2)
(f) (1, 2)

If the desks in *your* room are in rows and columns, what are your co-ordinates? (Imagine you're in the teacher's place; count from the left, then from the front.)

Choose some other people in your class and write down their names and co-ordinates. Check with them to see if they agree!

(2,3) OR NOT (2,3)? THAT IS THE QUESTION!

Now, Class 2M thought that everyone should have a chance to sit by the window! So at the end of every half-term and at the end of every term, the window column moved to the door side of the room and the other columns moved to the right one place.

Example: after one move, Michael is now (4, 3) and Brian is (1, 2).

Give the new co-ordinates for
(g) Susan **(h)** Margaret **(i)** Richard

Who is in the following seats after two moves?
(j) (1, 4) **(k)** (3, 3) **(l)** (4, 2)

(m) Where should Darren be at the beginning of the third term?

When everyone was back where they started, the girls at the front complained that they wanted to sit further back! So... every half-term, starting with the *beginning* of the third term, each row moved back one and the back row came to the front. Don't forget – the columns are still moving too!

Example: at the beginning of the third term Helen is (2, 2).

Give the co-ordinates for these pupils at the end of the third term:
(n) Nicola **(o)** Lucinda **(p)** Everton

P.S. The teacher was so confused by now that he left!

20 The badminton tournament

★★

The diagram shows the plan of a badminton court.
Below are the dimensions in feet and inches. Use
the conversion chart to rewrite these in metric
form.

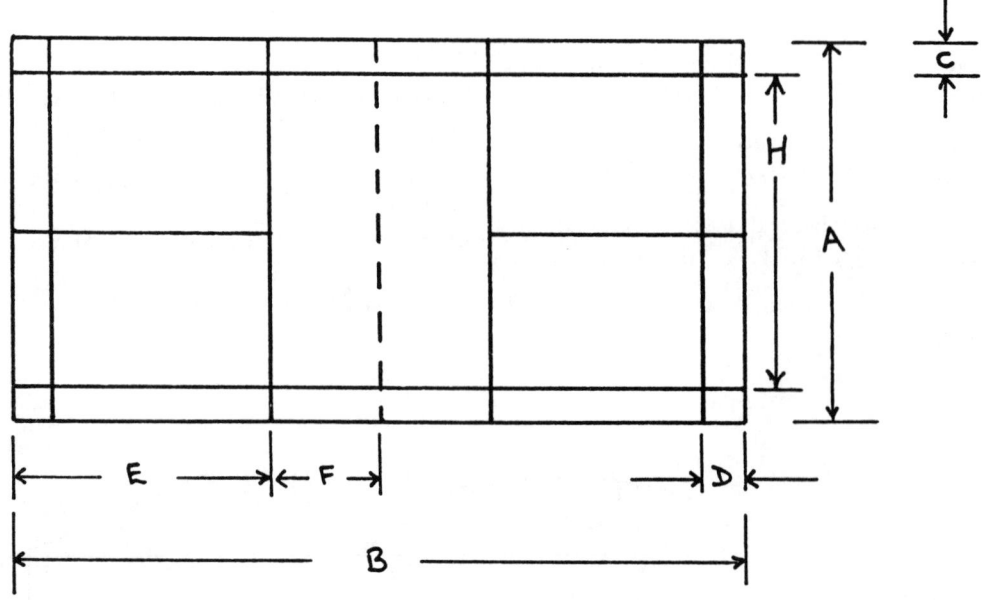

		Feet	Inches			Feet	Inches
A	Doubles Width	20 ft	0 ins	**E**	Singles Service Court	15	6
B	Length	44	0	**F**	Net Area	6	6
C	Tramline	1	6	**G**	Height of Net	5	0
D	Back Service Tramline	2	6	**H**	Singles Width	?	?

Conversion chart	
Feet	**Metres**
1	0.30
2	0.61
3	0.91
4	1.22
5	1.52
10	3.05
20	6.10
30	9.14
40	12.19
50	15.24

In a school tournament between 4 Houses, each House entered a boy and a girl for Singles in each Year, and a mixed pair for Doubles in each Year. Years 1–5 took part.

Below are the results chart and points table for Third Year Boys' Singles. (Note: 2 points for a win, and 1 point bonus for each 10 game points scored.)

House	Green	Red	Yellow
Blue	15–6	13–15	15–1
Green		15–8	15–2
Red			11–15

Examples:
Blue beat Green 15–6
Yellow beat Red 15–11

| House | Played | Won | Game points | | Bonus | Total points |
			For	Against		
Blue	3	2	43	22	4	8
Green	3	2	36	25	3	7
Red	3	1	34	43	3	5
Yellow	3	1	18	41	1	3

Now make a results chart and points table for the following matches:

First Year Doubles

Blue	7–15	Green
Red	10–15	Yellow
Blue	15–10	Red
Green	5–15	Yellow
Blue	2–15	Yellow
Green	9–15	Red

In each competition there are 6 games to be played. How many games are played in the whole tournament? If each game lasts an average of 12 minutes, how long does the Sports Master have to umpire for?!

21 Stick at it!

The designs in this exercise are made with match-sticks. Try to solve the problems just by thinking and drawing!

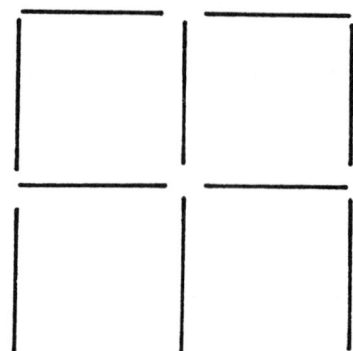

In the diagram on the right
(a) how many squares are there?
(b) remove two matches to leave only two squares.
(c) move three matches to leave three identical squares.

An 'equilateral' triangle has all 3 sides the same length. Can you make 6 equilateral triangles using only 12 matches? If you can do that, now remove 4 matches to leave 3 triangles.

The angel fish about to swim off the page is made with 8 matches. Move just 3 matches and make it swim the other way!

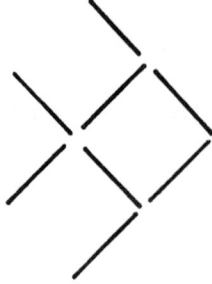

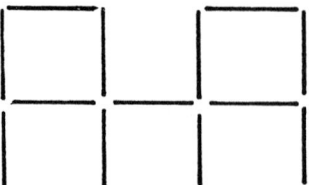

Move one of these matches and leave nothing!

Take the cherry out of the glass by moving only 2 matches. You mustn't touch the cherry!

Move only three matches in this design, to leave 4 equal squares.

22 Bearing up

The angles between directions are called 'bearings' and are measured in degrees on a compass. One complete turn is 360 degrees. Bearings are always measured *clockwise*, from North.

Example: East is a bearing of 90° and is $\frac{1}{4}$ of a full turn.

Work out the bearings and fractions of a full turn for these directions:

1. West
2. South
3. North-east
4. South-east
5. North-west
6. North-north-east
7. West-south-west
8. West-north-west
9. South-south-east
10. North

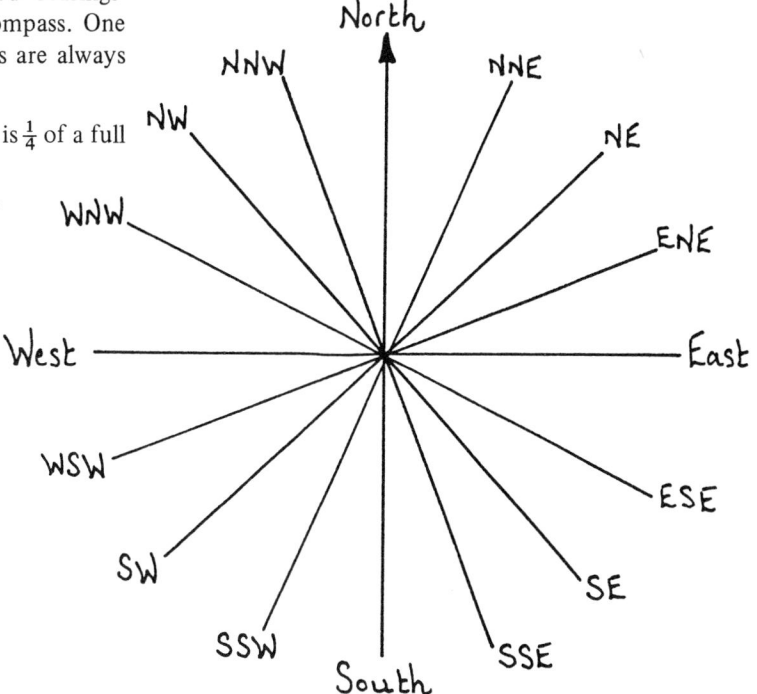

What are the bearings from the first to the second direction in these pairs?

11. E $\longrightarrow$ S
12. S $\longrightarrow$ N
13. E $\longrightarrow$ N
14. NE $\longrightarrow$ SE
15. NE $\longrightarrow$ W
16. WSW $\longrightarrow$ WNW
17. SSE $\longrightarrow$ NNW
18. W $\longrightarrow$ SW
19. ENE $\longrightarrow$ NE
20. NNE $\longrightarrow$ ENE

23 Number puzzles 1
★★

In a pentacle (5-pointed star) the number 72 is very significant – ask your maths teacher why! Try to fit numbers in the spaces on the pentacle shown, so that every line adds up to 72.

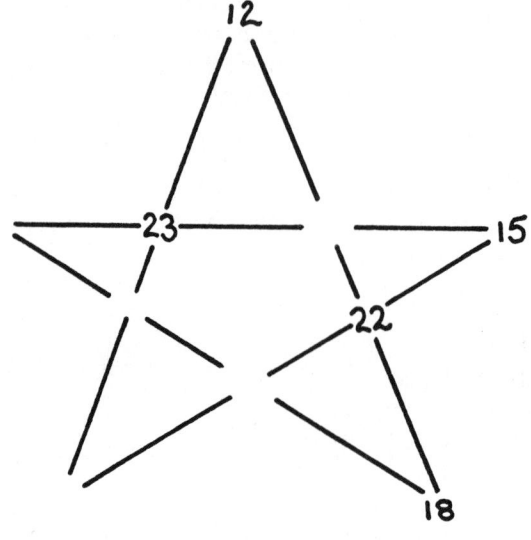

When my sister was 36, my father was 63. How old was she when she was exactly one quarter of his age? How old will he have to be for her to reach three quarters of his age? (Hint: it's over 100!)

Can you find the missing number:
2, 6, 15, 34, ?, 152.

A rat was crawling up a ship's rope which was 30 feet long. It climbed up 3 feet a minute but then slipped back 2 feet with exhaustion! How long did it take to reach the ship?

24 Get set...
★★

A 'set' is any group of things that have something in common. The things are called the 'elements' of the set and what they have in common is the set 'characteristic'.

Example: elements: dog, horse, cat, teacher
characteristic: animals
Example: elements: dog, cat, table, chair
characteristic: four legs

When a class arrived for maths, 20 pupils had forgotten their rulers and 17 had forgotten their pencils. When they were told to go and get them, 24 pupils went out! How many had forgotten *both* pencil and ruler?

dog	sea	brick	motorboat
table	oil	paper	sand
red	gull	green	'plane
grass	sun	windmill	pen
venus	radio	tooth	watch
vegetable	horse	train	money
water	chair	finger	lemonade
milk	blue	earth	teacher
fire	sky	stone	leaf
bed	tree	card	flower
ruler	flag	abacus	wine
calculator	cat	eye	swan
planet	car	owl	moon
computer	TV	pencil	brown

Now, how many different sets can you find from the list above? Each set should have at least 4 elements, and each characteristic may only be used once.
(1–15 not trying! 16–30 average; 31–50 very good; more than 50 - brilliant!)

25 Number puzzles 2

★ ★

Can you put all the numbers 1-9 in the boxes on the right so that all four sums are right?

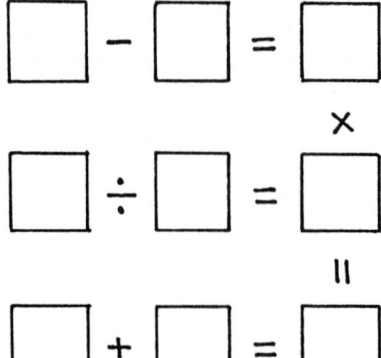

On the right is the key for a variation of 'pigpen code'.

Example:

C is ▢

D is ▢

A B C 1	D E F 2	G H I 3
J K L 4	M N O 5	P Q R 6
S T U 7	V W X 8	Y Z ? 9

Can you decode the message on the right? When you've done it, try to write your own coded message.

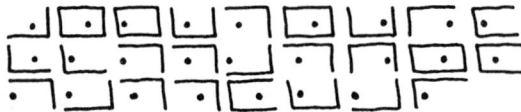

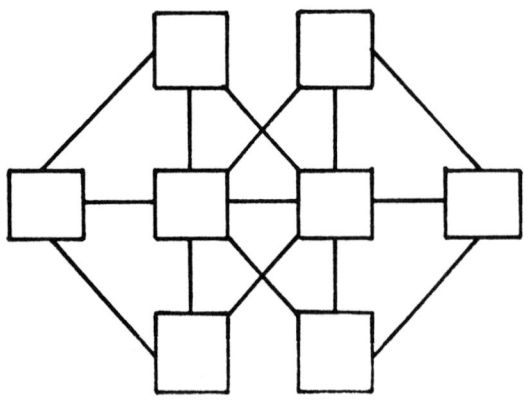

Can you put all the numbers 1-8 in the boxes so that a line never connects two consecutive numbers?

100 people were asked which BBC programmes they had watched the night before. 73 had watched BBC1, 28 had watched BBC2, and 12 had watched both. How many didn't watch either?

26 Logic 2
★ ★

Try to find the missing items in these series:

1. M, T, W, T, F, S, _
2. J, F, M, A, _, J, J, A, _, _, N, D
3. 1, 1, 2, 3, 5, 8, _, 21, _, _, _
4. 1, 4, 9, 16, _, _, 49, 64, _, 100
5. 1, 3, 6, 10, _, 21, 28, _, _, 55

6. A, T, D, Q, G, _, _, K, M
7. A, AB, ABD, ABDG, _
8. YZ, TS, MN, HG, _
9. 1, 2, 5, 11, 21, 36, 57, _, 121
10. O   A   B   S   C   M   S   C   P   –   D

(only try the last one if you're very clever!)

Try to draw the missing shapes in these series:

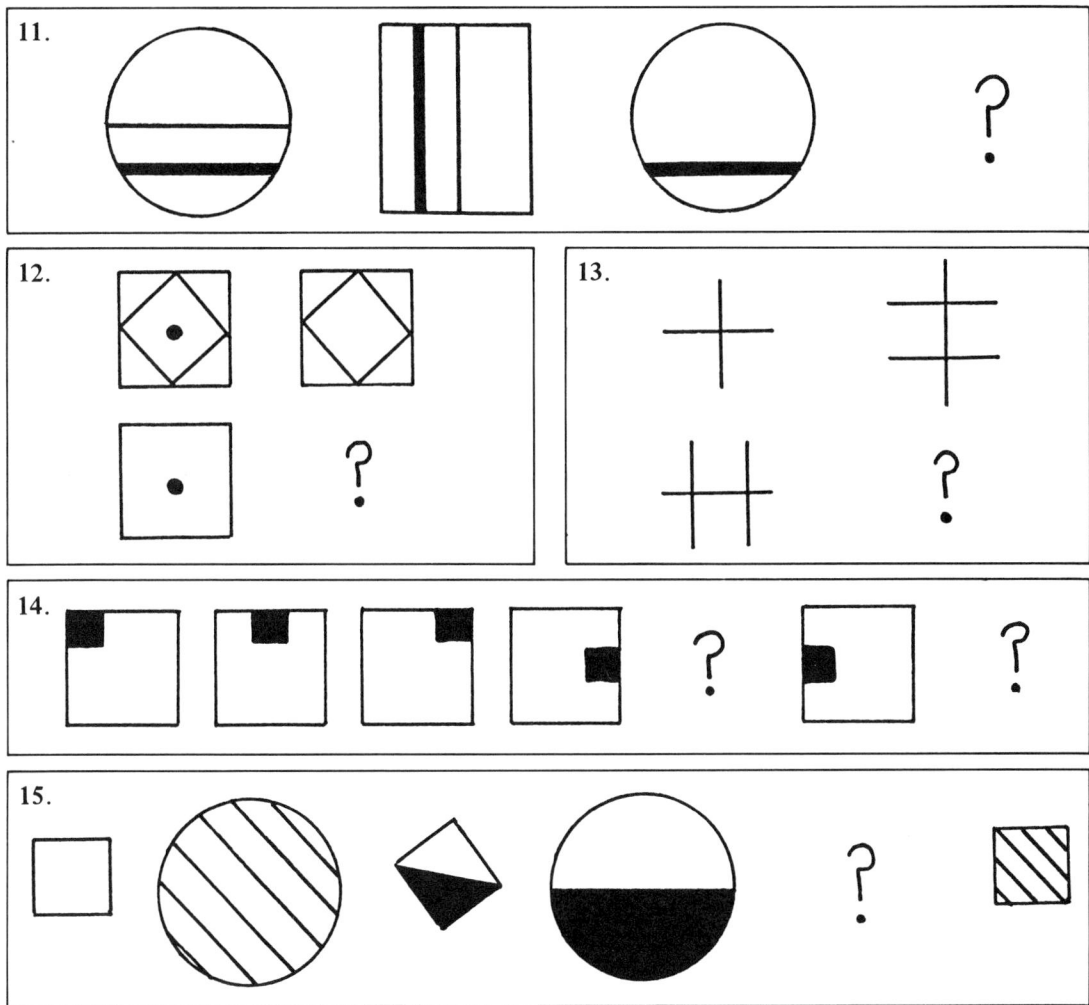

Can you make up some more puzzles like these?

27 Dicey problems

In a certain game each player throws 4 dice. He can add or subtract any of the scores to or from each other, and the winner is the player whose total is nearest to ten.

Example: I throw 2, 4, 6, 3.
(a) $4 + 6 + 3 - 2 = 11$
or I could arrange them:
(b) $4 + 6 + 2 - 3 = 9$

Both arrangements are 1 away from 10 – I can't get closer!

The throws in a particular game are shown on the right. Work out each player's best total and decide who won and who was second. Who would have won if the object was to get the *lowest* score by adding and subtracting?

Anne	1,	4,	6,	6
Brian	5,	5,	6,	2
Chris	2,	2,	3,	5
Danny	6,	6,	6,	6
Elaine	3,	4,	5,	6
Frank	1,	2,	3,	2
Georgina	1,	5,	1,	1
Henry	2,	2,	2,	1

In another game each player throws 3 dice. He must add 2 numbers and multiply by the third, trying to get the highest total. Who won this game?

Who would have won if they had to subtract one number from another then multiply by the third to get the *lowest* total?

Anne	2,	3,	5
Brian	4,	4,	4
Chris	3,	3,	5
Danny	1,	4,	6
Elaine	4,	5,	3
Frank	1,	5,	6
Georgina	2,	2,	6
Henry	2,	5,	5

Try to make up a game like those on the opposite page, using 5 dice. Invent some scores for, say, four players and work out who won. Then see if one of your friends can work it out correctly!

Here's a good investigation. You will need two dice, or two hexagonal spinners, or you can just use two hexagonal pencils. If you can't get these now, try it at home later. Remember the story of the tortoise and the hare? Mark the sides of one of your dice or spinners in the normal way from 1 to 6; this one is for the tortoise. Now mark the sides of the other for the hare: 0, 6, 0, 6, 0, 9. (The hare runs in leaps and bounds, with rests between!)

Roll the dice alternately and keep a running total for each. The winner is the first to reach 50. Who usually wins?

In the game of Crap you have to score 7 with two dice. Find out what the chances of doing this are by listing all the possible scores and seeing how many add up to 7.

How many spots are there on a die? What is the *average* score when you roll a die? What, then, is the average when you roll *two* dice?

28 Milko!

A milkman wrote the following table to remind himself what to take into a block of flats each day.

	Flat							
	A	**B**	**C**	**D**	**E**	**F**	**G**	**H**
Silver top	2	3	0	0	2	2	4	0
Red top	0	0	3	3	2	0	0	1
Gold top	0	1	0	0	2	2	0	1
Eggs	6	0	0	0	6	0	0	12
Loaves	1	1	1	0	0	0	0	1

(a) How many bottles of each kind does he have to take each day? How many dozen eggs? How many loaves?

(b) If the order is the same every day, how many bottles of milk will he deliver in a week? How many eggs and loaves?

(c) Suppose silver top milk costs 23p a bottle, red top 24p and gold top 25p; eggs are 76p a dozen and loaves are 40p each.
Rewrite the table to show each flat's weekly total, then work out their bills.

(d) How much does the milkman collect from this block each week? What is his commission at 2% (2p per £)?

(e) One week, flats A, E and H go on holiday. By how much is the milkman's commission reduced this week?

(f) The milkman decides to stay in bed on Sunday mornings so he delivers Sunday's order the day before! Write a new Saturday table for him.

29 Points problems
★ ★

In an athletics match between 4 schools there were 10 points for a win, 7 points for being second, 4 points for third, and 1 point for fourth place. The results were as follows:

Event	1st	2nd	3rd	4th
100 m	Acacia Road	Belsize Street	Charles II	Deans School
1500 m	Belsize Street	Charles II	Acacia Road	Deans School
Long jump	Deans School	Belsize Street	Acacia Road	Charles II
High jump	Deans School	Charles II	Acacia Road	Belsize Street
Javelin	Deans School	Belsize Street	Acacia Road	Charles II
Relay	Charles II	Acacia Road	Deans School	Belsize Street

Copy and complete the following table:

School	1sts	2nds	3rds	4ths
Acacia Road	1	1	4	0
Belsize Street				
Charles II				
Deans School				

Use your table to find out who won, who was second, third and fourth. Why did the final result depend on the Relay?

Work out what the result would have been if there had been 4 points for a win, 3 points for second, 2 points for third and 1 point for being fourth. Does it make any difference?

Invent a different score system and see if that makes any difference to the final result.

30 What I did in the holiday

★★

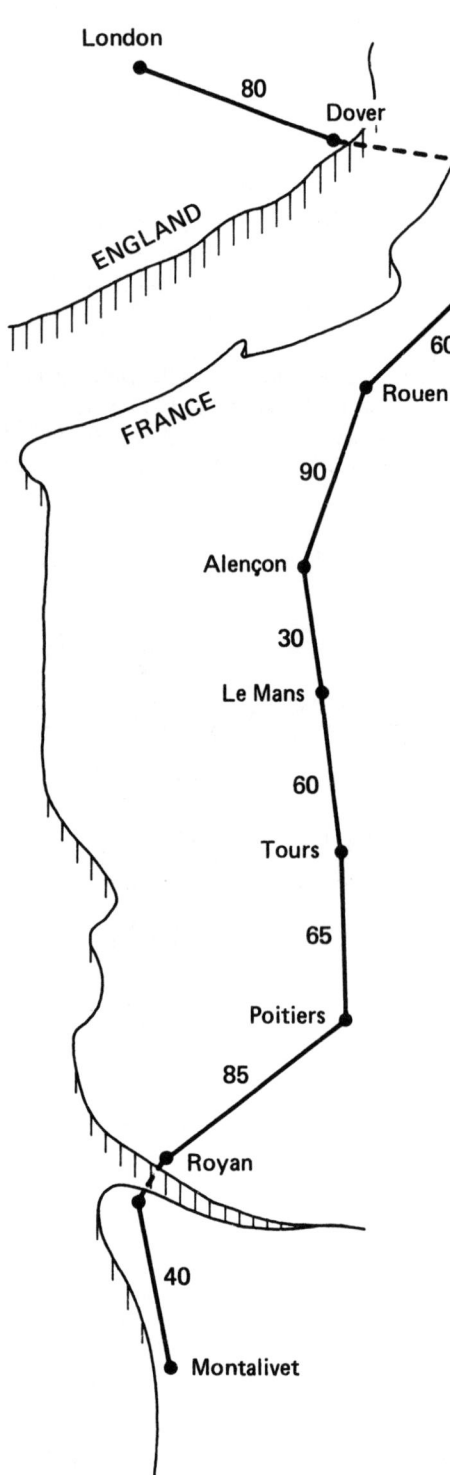

_ _ _ _ _ _ _ _ _ shows a ferry crossing.

The map shows the route I drove on holiday from London down to Montalivet in south-west France. The distances shown are _miles_.

What are these distances? Ignore the sea crossings.

(a) London–Dover
(b) Boulogne–Le Mans
(c) Le Mans–Poitiers
(d) Poitiers–Montalivet
(e) London–Montalivet

When I left home the car's mileometer read 24 650 miles. What did it read at:

(f) Dover
(g) Boulogne
(h) Rouen
(i) Tours
(j) Montalivet?

What are these distances in km?

(k) London–Dover
(l) Boulogne–Abbeville
(m) Alençon–Le Mans
(n) Royan–Montalivet
(o) London–Montalivet

The table below shows the times I arrived at and left each town in France.

Town	Arrived	Left
Boulogne	–	8.15 a.m.
Abbeville	9.30 a.m.	9.30 a.m.
Rouen	11.15 a.m.	11.45 a.m.
Alençon	2.00 p.m.	2.45 p.m.
Le Mans	3.30 p.m.	10.00 a.m.
Tours	11.45 a.m.	1.00 p.m.
Poitiers	3.00 p.m.	3.45 p.m.
Montalivet	6.00 p.m.	–

(p) Where did I have lunch on the first day?
(q) Where did I stay overnight?
(r) Where did I have lunch on the second day?
(s) How long did it take to get from Boulogne to Rouen?
(t) How long did it take to get from Boulogne to Le Mans?
(u) How long did I spend *actually driving* in France?
(v) What was the total distance driven in France, in miles?
(w) Divide **v** by **u** to find my average speed in France.
(x) If 5 miles = 8 km, what was this average speed in kmph?

31 The right way
★★

The path of a journey can be drawn as shown below, where 1 cm represents 1 km.

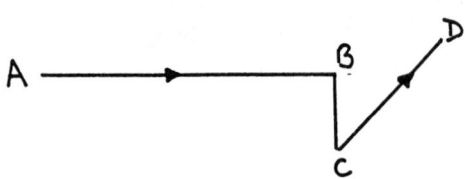

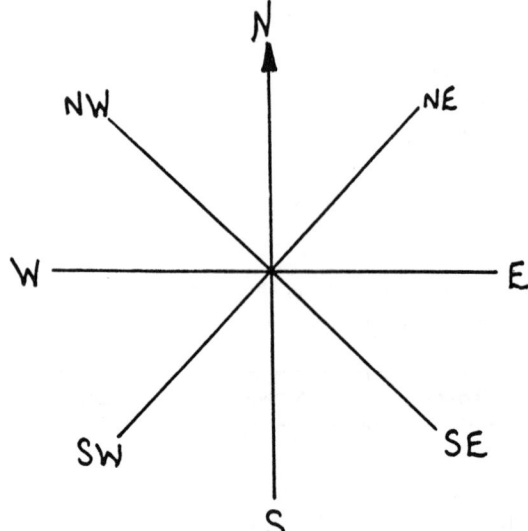

This journey can be described as: 'From A go 4 km east to B, then go south 1 km to C, and finally north-east 2 km to D.'

Draw careful diagrams to represent these journeys:

1. From A go 4 km east to B, then 1 km north to C, then 1 km west to D.
2. From A go 5 km south to B, then 2 km east to C, then 3 km north-east to D.
3. From A go 2 km north to B, then 7 km west to C, then 3 km south to D, then 5 km south-east to E.
4. From A go 4 km north-west to B, then 3 km south-west to C, then 2 km south to D, then 9 km east to E, then 1 km north-east to F.

Describe the following journeys in words, giving distances and directions:

5.

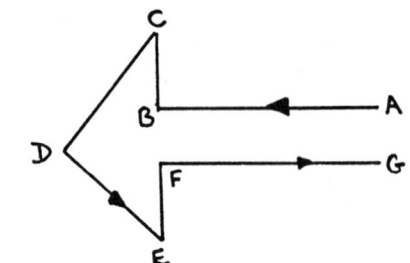

6.

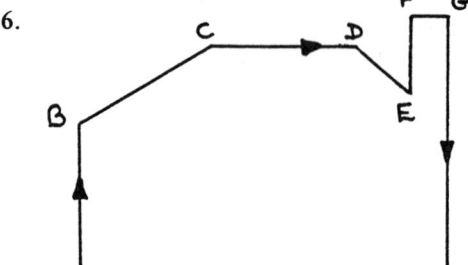

Make up some journeys of your own like these and try them on your neighbour.

32 Equation search 2

Search for sets of numbers that can be put together to make equations in horizontal, vertical or diagonal lines.

Example:

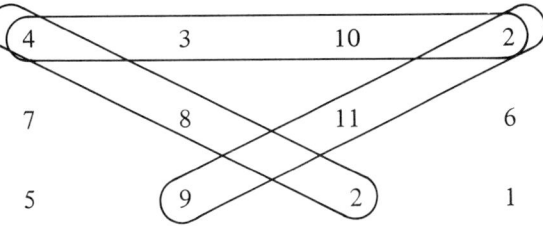

The numbers ringed form equations because

$4 \times 3 = 10 + 2$

$4 = 8 \div 2$

$9 = 11 - 2$

Copy the grid below and find as many equations in this way as you can. There are more than *forty*!

3	2	6	7	5	11
4	3	9	9	1	2
2	1	0	3	20	2
6	6	7	3	5	10
8	2	7	9	4	10

33 Logic puzzles 6

★ ★ ★

Tricks with digits: can you make 37 with five threes?!

Answer: $33 + 3 + \frac{3}{3}$

Now try these:
use 6 nines to make 100
use 6 sixes to make 120
use 4 fours to make 60
use 4 nines to make 20

Can you draw a diagram showing how 12 coins can be arranged in a square with 5 on each side?!
Then try to arrange 12 coins so that there are 6 rows with 4 coins in each row. . . .

A goods train with an engine and 5 trucks stopped at a small station where the siding was only big enough for 3 trucks (or 2 trucks and an engine). How can a passenger train get through?

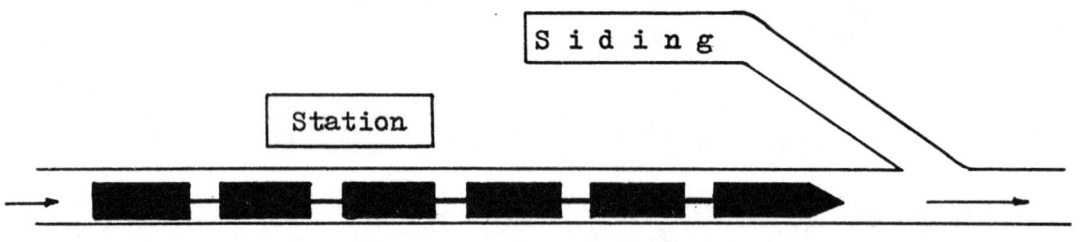

Six children found some money, an amount between £1 and £2. They divided it up equally but there was 5p left over. Then 2 of their friends arrived so they shared the money equally between the 8 of them – but there was still 5p left! One of them asked his older brother's advice, but *he* wanted a share too. So they shared it out again between the 9 of them. . . Guess what? Still 5p left over! How much had they found?

34 X-number puzzle 3

★ ★ ★

Copy the grid carefully first.

Clues across

(a) 456 − 333

(c) CXCIV

(f) May days

(g) 1 + 6 × 16

(i) 5^2 × 11

(j) January

(l) almost an emergency!

(n) 62 × 6

(p) one ninth of **d** down

Clues down

(b) 67 + 143 + **j** across

(d) a real emergency!

(e) five to nine

(f) palindromic (but nothing in the middle!)

(h) 181 × 4

(k) second prime after 11

(m) **g** across + 95 − 101

35 X-number puzzle 4

★ ★ ★

This puzzle is rather harder than the earlier ones because all the clues are cryptic. So think carefully and good luck! Copy the grid first.

Clues across

(a) one to four times
(c) reverse of **a** across!
(d) these clues are grossly unfair
(e) is 3 backward?
(f) two days of hours
(h) this should take half a minute
(i) this one, five sixths of a minute!
(j) one less than a prime whose digits add up to 11
(l) a prime!
(m) 192, in an upside down sort of way
(n) bakers have dozens of these
(p) 3 seconds after **i** across

Clues down

(b) five eighties and twos
(c) two lots of three twos!
(e) three, and nothing else but two
(g) days from 30th June 1989 until Fireworks Day 1991, inclusive
(k) in a daze for two years and three long months
(l) three weeks of May run together

36 Playing with numbers
★ ★ ★

Five people are having a darts match. They each start at 501 and they subtract their score at each turn to reduce this to 0, ending with a 'double'.

Example: if I've scored 470, I need 31 more. This requires at least 2 darts: I could get 1 + dbl 15, or 3 + dbl 14, etc.

The five players' scores so far are given below. Work out what they need to finish and how many darts it will take. Give an example of how they could do it. Who is most likely to win?

Anne 15, 19, 17, dbl 14, 20, 20, trb 20, 3, 4, 18, trb 11, trb 20, trb 20, trb 20, trb 19 . . . .

Brian 25, 25, dbl 17, 9, 19, trb 18, dbl 13, 1, 1, trb 20, trb 20, trb 20, 50, trb 20, 15 . . . . .

Chris 19, 19, dbl 19, dbl 20, 50, trb 17, 18, trb 9, 16, trb 20, 19, dbl 18, dbl 18, trb 20, 5 . . . .

Dave trb 20, trb 20, trb 20, 1, dbl 20, 18, 50, 25, 25, trb 19, 14, 2, 2, dbl 20, trb 13 . . . . . .

Eddie dbl 6, 19, 19, dbl 17, 15, trb 19, 3, 18, dbl 15, dbl 13, trb 20, dbl 18, 17, 12, 18 . . . .

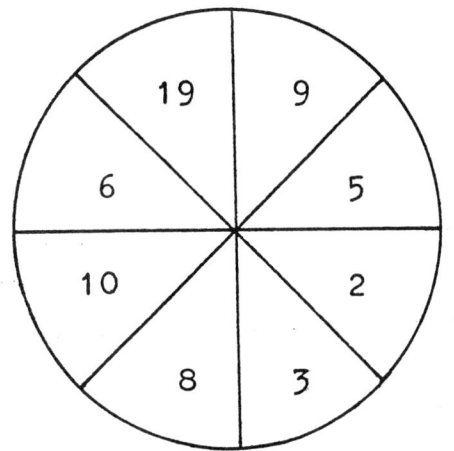

With the dartboard on the left, how many different ways are there to score exactly 20 with 4 darts? You can't use the same 4 numbers twice in a different order!

37 Working with numbers

A factory employs 400 people. Half of them work 7 hours a day, 5 days a week. The other half work 8 hours a day for 5 days a week. Now, the union argues that everyone's hours should be cut, which would also create new jobs. . . Does this work in practice?

(a) In this factory, how many 'man-hours' are worked in a year of 48 weeks if there is no absence?
(Hint: people × hours per day × days per week × weeks)

(b) Suppose the working day is cut to 7 hours for everyone; how many man-hours are lost by this in a week?

(c) How many extra workers ('new jobs') would be needed to make up this lost time?

(d) What are the total wages for these extra people for a week at, say, £5 per hour?

(e) If the man-hours lost in a week were made up by people working *overtime* instead of by taking on new people, what would be the extra wages bill per week? (Overtime is paid at, say, 'time and a half' or £7.50 per hour.)

(f) Now, is it better to take on new workers or to increase overtime?

A small company proudly announced that its 'average' wage was £10 000 per year. Sounds good! On investigation, however, it was found that only the Supervisor earned this amount. The Secretary earned £8 500 and the Salesman £16 000, while the Manager earned £25 000! How much did the other six workers each earn? What percentage of the workforce actually earned less than the average wage? What percentage earned more? You can't always trust statistics!

38 More money trouble
★ ★ ★

In a competition the first prize of £500 was shared by 12 people. The second prize of £250 was shared by 6 people, and the third prize of £100 was shared by 2 people. How much did each winner receive? Which was the best prize?!

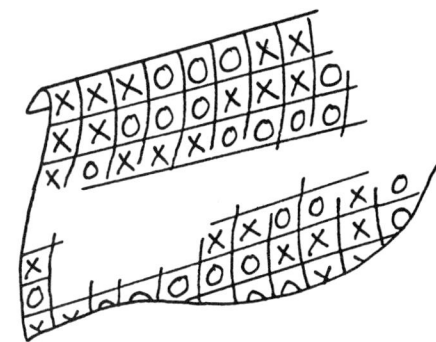

A pools syndicate of 19 people won £3801.90. How much did they get each?

A boy received £24 in presents for his birthday. He decided to spend one sixth on a disco, save a quarter, and buy some records with the rest (£4.75 each). How many records did he buy?

The prices at three local shops one week were as follows:

	Cereal	Eggs (doz.)	Soup	Margerine	Milk (pint)	Apples (per lb)	Bread
General Stores	84p	75p	22p	38p	25p	23p	45p
Supermarket	68p	72p	20p	29p	24p	28p	48p
N. E. Body & Co.	75p	75p	21p	30p	25p	20p	48p

PKT CEREAL
DOZ EGGS
SOUP - 3 PKTS
MARGE.
MILK - 3 PTS
APPLES - 3 lbs
TWO LOAVES

Find the cost of the shopping list on the left for each shop.

The next week, the General Stores cut the price of each item by 2p as a sales promotion. How would the new bill compare?

A man earned £4 per hour with overtime paid at $1\frac{1}{2}$ times this.
(a) What is his basic pay for 40 hours (a normal week)?
(b) What is he paid for 44 hours?
(c) The union negotiates a rise of 30p per hour and a cut of 2 hours in the normal working week. What did he earn in a normal week then?
(d) If he is offered night-shift work at £4.70 per hour for a 35 hour week, should he accept?

39 (Magic) X-number puzzle 5
★ ★ ★

The letters **a – i** stand for numbers which can be found by solving these clues:

$5a - 20 = 0$
$2b + 1 = b + 10$
$c = b - a - 3$
$3d = b$
$3e = a + b + c$
$f + 2f - 4f = -7$
$g + h = b$
$h = ab - ef$
$i = f - h$

a	b	c
d	e	f
g	h	i

Now draw the grid on the left and fill in the correct numbers.

There is something very unusual about this square of numbers. What is it?

What is **a + b + c**? What is **b + e + h**? What is **g + e + c**?

Add up all the numbers and divide by 3. What is special about the answer?

Now divide the total of the numbers by the number of numbers (in other words, find the average). Where is this number in the 'magic square'?

Using your answers to these questions, try to make another magic square with the numbers in different positions.

In a 5 × 5 magic square, what would be the total of each row? What would the middle number be?

40 Talking machines
★ ★ ★

Some numbers in digital displays (like the display on a calculator) look like letters when viewed upside down. Below are the numbers 0–9 with the letters they resemble:

0	1	2	3	4	5	6	7	8	9
O	I	z	E	h	s	g	L	B	b

See how many words you can make from these letters, and write the corresponding upside-down number.

Target:

20 – average

30 – good

40 – excellent

Examples:

bell = 7739

size = 3215

slob = 9075

(Remember: read the word backwards!)

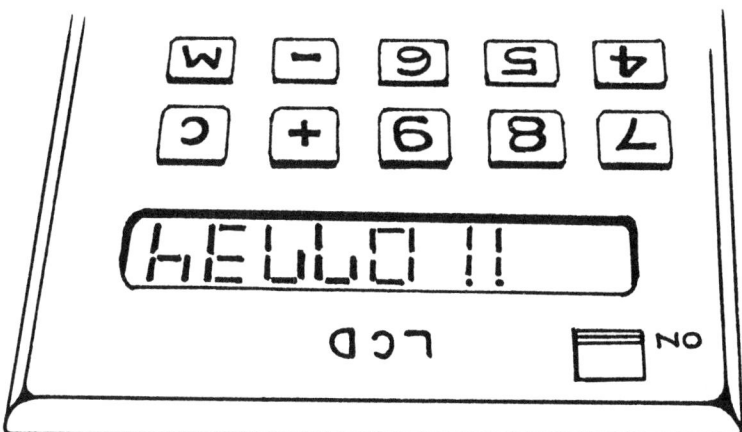

Work out these sums and find the word the answer makes:

115 469 × 5	Clue: nuts have them
80 × 9 + 13	Clue: shocking fish!
31 × 11 × 11	Clue: in the middle of water
1107 × 50 + 29	Clue: priests often do it
440 × 7	Clue: this sum will be instrumental in improving your maths!

Try to write a short sentence with 'digital words'. Then write the words as numbers and make up clues for them. Try them out on your friends.

41 Logic puzzles 7

★ ★ ★

Here's an old problem! A man has to take a wolf, a goat and a cabbage across a river but his boat is only big enough for himself and *one* other item. If he takes the cabbage, the wolf will eat the goat! If he takes the wolf, the goat will eat the cabbage! Assuming that wolves don't like cabbage and animals can't row boats, how did the man get them all across safely?

A ship stands in harbour with a rope ladder over its side. The ladder has 20 rungs, each 1 foot apart, and the bottom rung just touches the water. As the tide comes in, the water rises 6 inches per hour. . . How long will it take for 3 rungs of the ladder to be covered by water?

Can you draw a diagram to show how 16 coins can be arranged so that there are 10 rows which contain 4 coins each? Now, can 9 coins be arranged so that there are 10 rows which contain 3 coins each?

A bus leaves London for Brighton at 2 o'clock, travelling at an average speed of 30 m.p.h. At the same time a cyclist leaves Brighton for London, travelling at an average speed of 10 m.p.h. Assume it is 60 miles from London to Brighton. Now, when the bus and the cyclist pass each other, which one will be furthest from London?

Which is worth more: half a pound of gold sovereigns or a pound of gold half-sovereigns?!

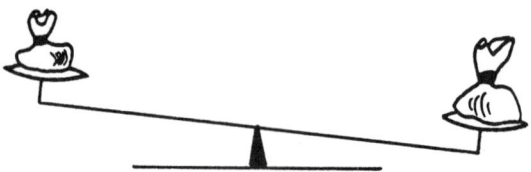

42 Logic puzzles 8
★ ★ ★

If 7 + 8 = 3, and 10 + 6 = 4, and 2 nines make 6. . .
what is 6 + 7 ? (You'll need time to do this one!)

How many squares can you find in the diagram
below?

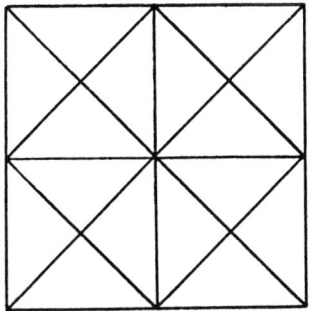

Draw a diagram to show how you could place ten
chairs along the walls of a square dance hall so
that there are an equal number of chairs along
each of the four walls!

A prisoner is in the cell marked X and all the doors
are open! He only has to walk out. . . BUT if he
enters the last cell without having gone through
all the others, *or* if he enters any cell more than
once, electronic sensors under the floor will lock
all the doors! How can he escape?

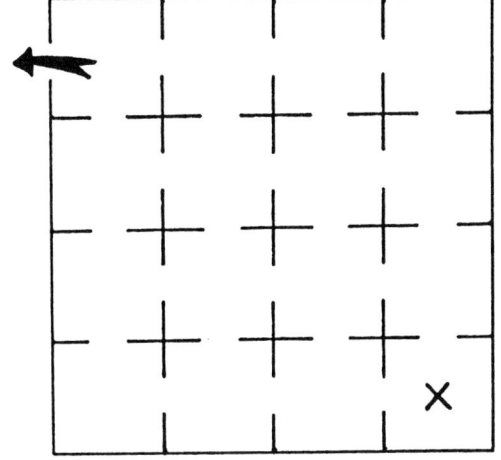

Three escaping prisoners come to a deep, fast river
with no bridge. However, there are two small boys
with a boat. But the boat is only big enough for
one man or the two boys. How can the men get
across? If each one-way crossing takes 5 minutes
and the police are 50 minutes behind, can they all
escape in time?

43 Logic 3
★ ★ ★

Put these actions in a logical order. Then see if you can find a different but equally correct logical order.

Driving a car away

(a) put the car in gear
(b) release the brake
(c) switch on the engine
(d) engage gear and move off
(e) get into the car
(f) put the key in the ignition
(g) unlock the car door
(h) check the gears are in neutral
(i) make sure it's safe to move off, and signal

Wallpapering a room

(a) buy the paper and paste
(b) strip off the old paper
(c) clean the brushes
(d) choose the new paper
(e) decide how much paper to buy
(f) calculate the area of the walls
(g) measure the walls
(h) get out the brushes
(i) decide how much paste is needed
(j) put up the new paper

Try several numbers in the flow chart below and make a note of how many 'loops' you have to do before escaping. What sort of number must you choose in order to escape quickly? (**Hint**: try 5, 6, 7, 8, and 9 first, then 15, 16, 17, 18 and so on.)

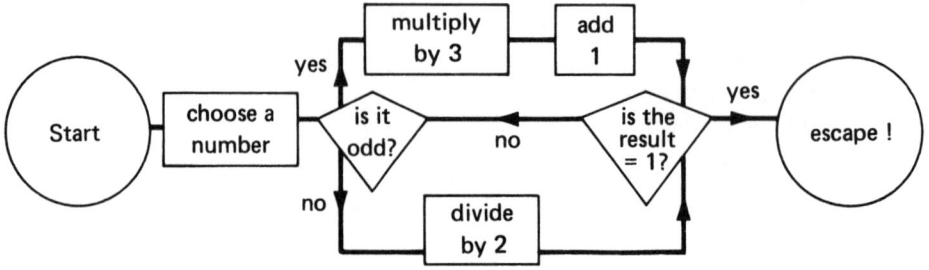

44 Stick at it again

The diagrams here were made with matchsticks. Try to solve the problems just by thinking and drawing!

The diagram on the right shows a mathematician's house and garden – both square! How can he split the garden into 5 plots, all identical in size *and* shape? (You'll need to use 10 'matchsticks' to divide the garden up.)

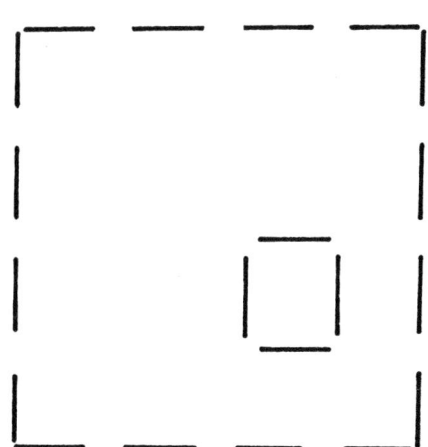

On the left is a similar problem – very tricky! How can you divide this garden up into 4 plots of equal area, so that each plot touches all 3 of the others?

Can you make 3 squares with just 10 'matchsticks'?

How many squares are there in the diagram on the right?

Now remove just two 'matchsticks' so that only three squares are left. . .

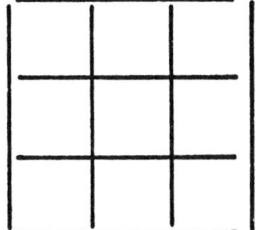

Make a square from 6 matches by breaking the rules. . !

45 How long to go?
★ ★ ★

Below are the plans of a football pitch and tennis court. Use the conversion chart to change the lengths into metres, and list your results in tables. Give all answers to 3 decimal places.

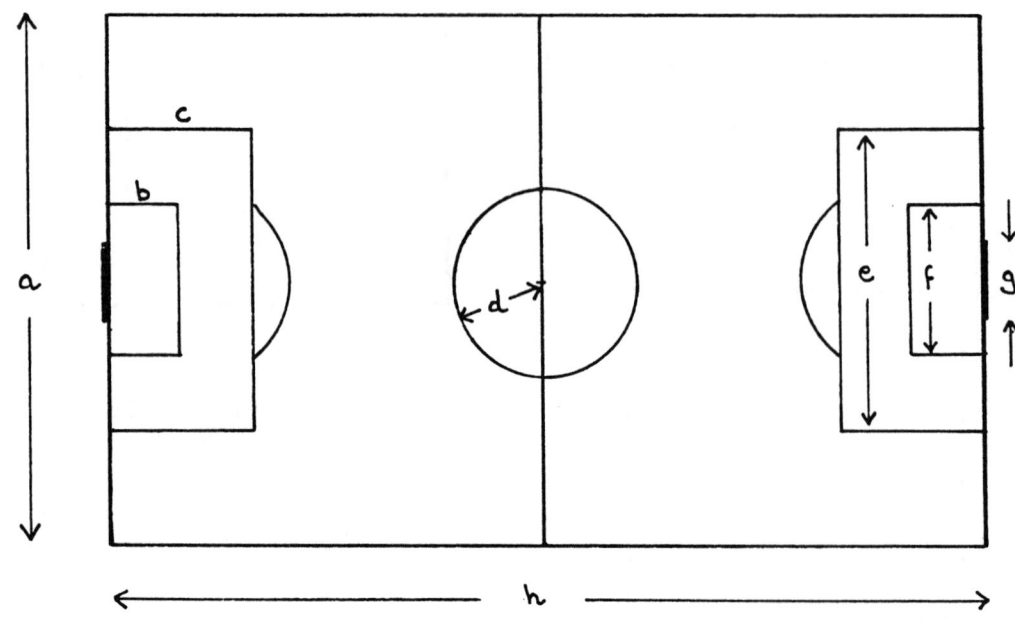

a = 228 ft   c = 54 ft   e = 132 ft   g = 24 ft
b = 18 ft   d = 30 ft   f = 60 ft   h = 348 ft

Feet	Metres
1	0.305
2	0.610
3	0.914
4	1.219
5	1.524
6	1.829
7	2.134
8	2.438
9	2.743
10	3.048

Feet	Metres
20	6.096
30	9.144
40	12.192
50	15.240
60	18.288
70	21.336
80	24.384
90	27.432
100	30.480

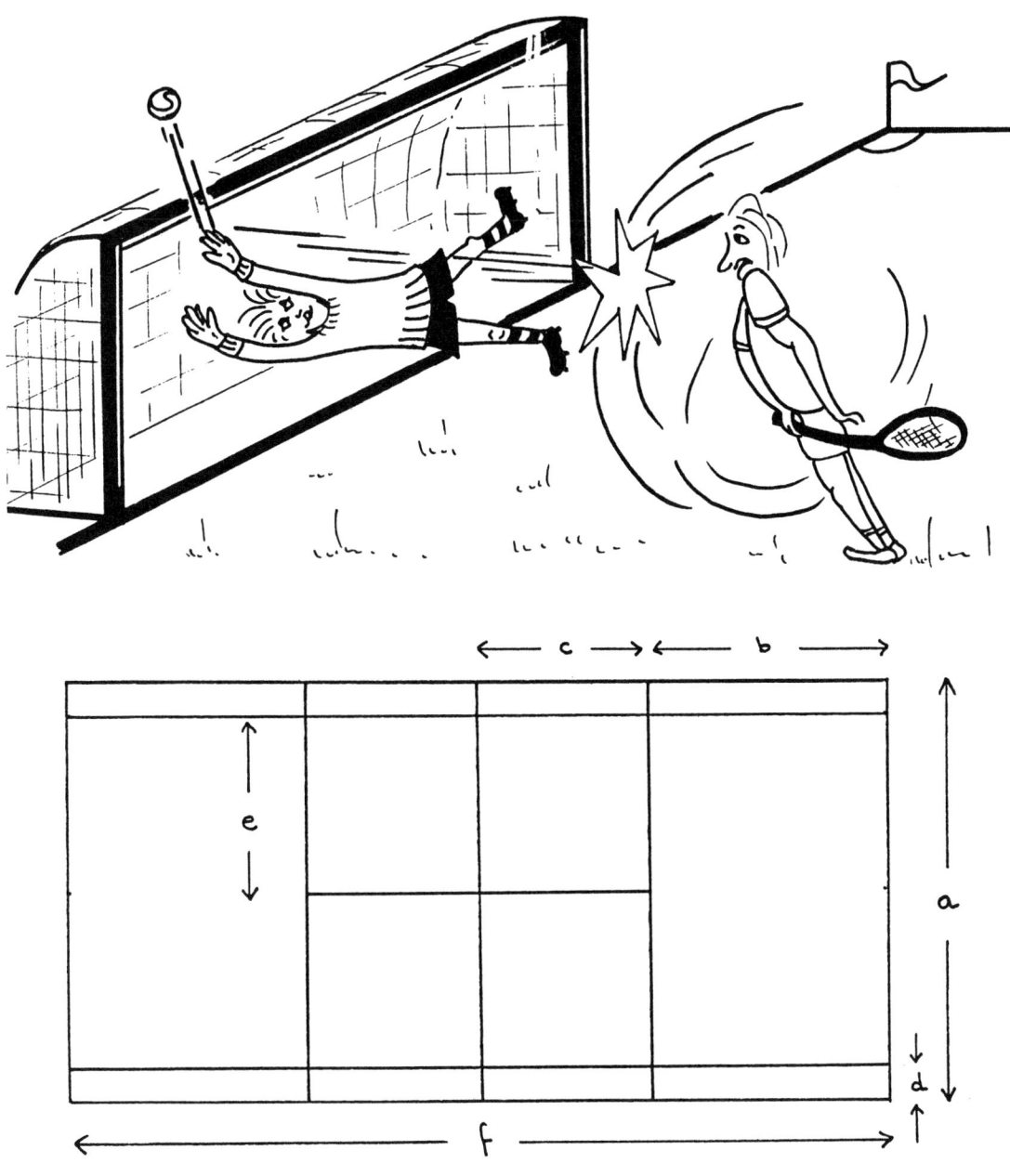

a = 36 ft   b = 18 ft   c = 21 ft
d = 4 ft 6 ins   e = 13 ft 6 ins   f = 78 ft

46 Elections

In a school election there were three candidates in each class, and pupils voted for them in order of preference. They got 3 points for being put first, 2 points for being second, and 1 point for being placed third.

In Class 4B the candidates were Cheryl, Andy and Len. The votes in Class 4B are on the right.

Find how many 1st places, 2nds and 3rds each candidate got, then work out their total points and decide who won.

Roughly, what percentage of the points did each candidate get?

Voter	1st	2nd	3rd
1	C	A	L
2	A	L	C
3	L	A	C
4	C	A	L
5	A	L	C
6	L	A	C
7	L	A	C
8	C	A	L
9	A	L	C
10	C	A	L
11	L	A	C
12	C	A	L
13	A	L	C
14	C	A	L
15	C	A	L
16	A	L	C
17	L	A	C
18	C	A	L
19	A	L	C
20	L	A	C

Now, suppose that only *first* choices counted... Who would have won?

What percentage of the votes would each candidate have got?

Compare these answers with your earlier ones. Can you comment on which system is fairer?

Now repeat your calculations for Class 3Y, where the voting was very strange!

The candidates were Chas, Alan and Linda.

What do you notice?

Voter	1st	2nd	3rd
1	A	L	C
2	A	L	C
3	C	L	A
4	C	L	A
5	C	L	A
6	A	L	C
7	A	L	C
8	L	C	A
9	A	L	C
10	C	L	A
11	A	L	C
12	L	C	A
13	A	L	C
14	C	L	A
15	A	L	C
16	A	L	C
17	C	L	A
18	C	L	A
19	C	L	A
20	A	L	C

In a Council election there were 10 000 voters using the same sort of 'preference' system as above. There were 10 points for being put first, 5 points for second, 3 points for 3rd, and 1 point for fourth. Who won?

Party	1st choice	2nd choice	3rd choice	4th choice
Conservative	4500	0	5250	250
Labour	1250	5000	3750	0
Alliance	4000	5000	1000	0
Independent	250	0	0	9750

47 Graphic codes

★ ★ ★

Here is a method of coding messages which uses a similar principle to that used in programming a computer to interpret or print out graphic images (pictures).

On a 5 × 5 grid as shown, block in squares to represent a letter or number – in this example, it's the letter E. (It's possible to represent more than one character on a grid at the same time.)

Now, for each row from top to bottom, write 0 if a square is blank and 1 if it is full.

In the example above:
1st row is   0   1   1   1   0
2nd row is   0   1   0   0   0   and so on.

Can you see how 'E' is coded as
0111001000011000100001110?

(If you're not sure, split the string into groups of 5 digits.)

Now, that's a lot of digits to represent one letter! But the code is very difficult to break since there's no clue in the string to the type of code used.

Can you code the word shown below?

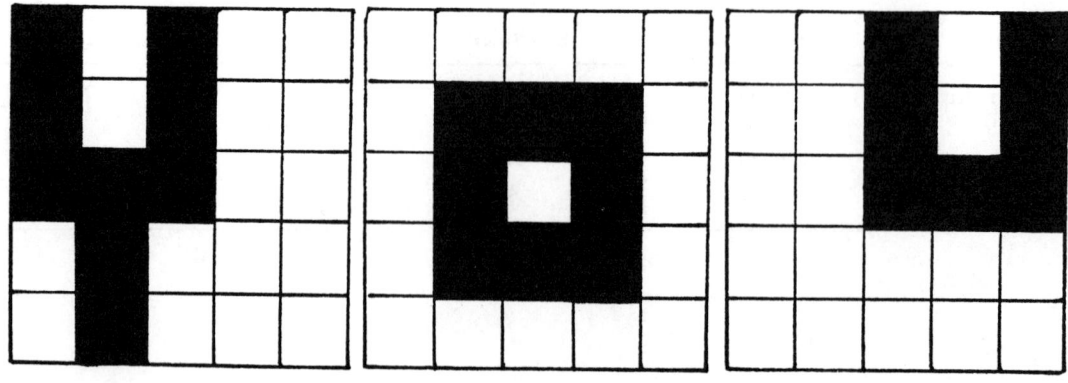

60

The last example shows that there are many variations possible for each character, and that the grid used could well be a different size. This makes the code more useful. Imagine, for example, that only you and the person you're sending the message to know what size of grid is being used! (Imagine that you have agreed to use a different size of grid for each alternate character. . !)

Try to decode the following message, written on 8 × 8 grids:

11101110100010101110111000000000
11011000101010001000100000000000
11100100100010101100111010001010
11100000000011100000010000000100
01000000010000000100110001010010
01000010000001000000100000011110

Hint: split the strings into groups of 8 digits. Every 8 groups will require one grid. . .

With a partner, decide a grid system to use and code a message to each other. When you've decoded each other's message correctly, ask someone else to try to decode one of them without knowing what system you've used!

(P.S. There *is* a way of telling what system might have been used, by counting the total number of digits in a message. Can you figure it out?)

48 Stumped!

In a one-day cricket international match between England and the West Indies a few years ago, England scored 270 for 8 wickets in the 60 overs they were allowed.

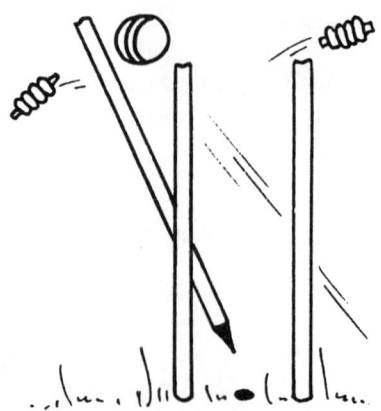

England's 'run-rate' was thus
270 ÷ 60 = 4.50 runs per over.
So this was the *average* run-rate the West Indies had to achieve to win the match.

Copy and complete the Table below which shows how they got on in their innings.

Overs gone	Score	Run-rate so far	Runs still needed	Overs left	Run-rate needed
5	21 for 1	4.2	250	55	4.6
10	61 for 1	6.1		50	
20	101 for 1		170		
25	126 for 4		145		4.1
30	150 for 5			30	
40	184 for 7	4.6	87		4.4
50	211 for 7			10	
55	241 for 8	4.4			
59	261 for 9	4.4			

By comparing the West Indies' 'run-rate so far,' and their 'run-rate needed', with England's average rate of 4.5, try to describe the progress of the West Indies' innings.

49 The beauty of numbers
★ ★ ★

You will probably need a calculator for this exercise.

On the next two pages, several fascinating arithmetic patterns are shown. In each case, try to write down the next two lines just by studying the pattern, and then check them by calculating.

The power of numbers

(a) $1^2 = 1$
$11^2 = 121$
$111^2 = 12321$

(b) $1^2 = 1$
$2^2 = 1 + 2 + 1$
$3^2 = 1 + 2 + 3 + 2 + 1$

(c) $1^3 = 1$
$2^3 = 3 + 5$
$3^3 = 7 + 9 + 11$

(d) $1^3 = 1^2$
$2^3 = (1 + 2)^2 - 1^2$
$3^3 = (1 + 2 + 3)^2 - (1 + 2)^2$

(e) $\dfrac{1 \times 1}{1} = 1$

$\dfrac{22 \times 22}{1 + 2 + 1} = 121$

$\dfrac{333 \times 333}{1 + 2 + 3 + 2 + 1} = 12321$

(f) $1 \times 2 \times 3 \times 4 + 1 = 5^2$
$2 \times 3 \times 4 \times 5 + 1 = 11^2$
$3 \times 4 \times 5 \times 6 + 1 = 19^2$

All at sixes and sevens

(g) $6 \times 7 \quad = 42$
$66 \times 67 \quad = 4422$
$666 \times 667 = 444222$

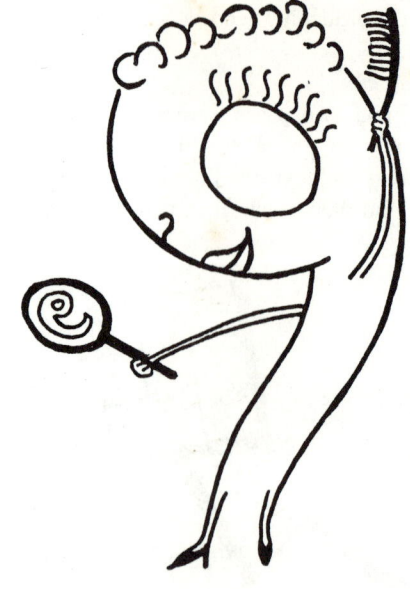

All at sixes and nines

(h) $9 \times 6 \quad = 54$
$99 \times 66 \quad = 6534$
$999 \times 666 = 665334$

Fractionally harder

(i) $1 \times \frac{1}{2} = 1 - \frac{1}{2}$ \qquad (check)
$2 \times \frac{2}{3} = 2 - \frac{2}{3}$
$3 \times \frac{3}{4} = 3 - \frac{3}{4}$

So is multiplying the same thing as subtracting?

(j) $4\frac{1}{2} \div 3 = 4\frac{1}{2} - 3$ \qquad (check)
$5\frac{1}{3} \div 4 = 5\frac{1}{3} - 4$
$6\frac{1}{4} \div 5 = 6\frac{1}{4} - 5$

So is dividing the same thing as subtracting?

Ad infinitum

Check these curious facts for yourself.
(k) any digit 'a' $\times 3 \times 37 \quad = aaa$

(l) 'a' $\times 3 \times 7 \times 11 \times 13 \times 37 = aaa\ aaa$

(m) 'a' $\times 9 \times 12\ 345\ 679 \qquad = aaa\ aaa\ aaa$